LIVING SYSTEMS

LIVING SYSTEMS

An Introductory Guide *to the Theories of* Humberto Maturana & Francisco Varela

Jane Cull

Copyright © 2000 Jane Cull

First edition, 2000
Second edition, 2004
Third edition, 2006
Fourth edition, 2013

ISBN 9781494356170

All rights reserved. No part of this publication may be reproduced, stored in a retrieval system, or transmitted in any form or by any means, electronic, mechanical, photocopying, recording or otherwise, without the prior permission of the copyright owner. Please purchase only authorized electronic editions, and do not participate in or encourage electronic piracy of copyrighted materials. Your support of the author's rights is appreciated.

Jane Cull, Life's Natural Solutions, lifesnaturalsolutions@gmail.com

Cover Design by Nada Backovic Designs
Typeset by Midland Typesetters, Australia

This book is dedicated to Dr Humberto Maturana whose explanatory gift in the biology of love and the biology of cognition is my passion. It is my desire that through this book your work becomes more readily available and understood to those within and outside the academic community.

CONTENTS

PREFACE	xi
ACKNOWLEDGEMENTS	xiii
CHAPTER 1 – THE ONTOLOGY OF THE OBSERVER	1
Scientific Explanations	10
Emotioning and the Two Explanatory Paths	11
CHAPTER 2 – THE OPERATION OF LIVING SYSTEMS	19
Defining Systems: Simple and Composite Unities	19
The Characterization of Living Systems: Autopoiesis	20
The Organization & Structure of Systems	22
How Living Systems Operate: Structural Determinism	24
The Nervous System	27
Function and Purpose	28
CHAPTER 3 – LIVING SYSTEMS AND THE ENVIRONMENT	29
Co-existence: Structural Congruence with the Medium	29

Interactions: Triggering, the Bodyhood & Behavior	32
Recurrent Interactions and Learning	35
Understanding Behavioral Change	37
Language	39
The Constitution of Human Beings	40

CHAPTER 4 – THE CONSTRUCTION OF WORLDS — 43

World Construction Overview	44
The Process of Objectification: Internal/External World	47
The Psychology of Objectivity-Without-Parenthesis: Patterns & Dynamics in Human Relations	50
Perceptual Differences	52
The Worldview of Objectivity-Without-Parenthesis	54
Separation: Perception	54
Separation: Behavior	55
Perceptual and Behavioral Uncertainty: Contradictions and Inconsistencies	58
The Worldview of Objectivity-in-Parenthesis	62
Relational Languaging: The Constitution of the Relational World	65
Moving Between the Two Worlds with Ease: Emotions	70

CHAPTER 5 – THE CONSTITUTION OF CULTURES — 73

Overview: The Patriarchal and the Matristic	73
The Patriarchal Culture as a Manner of Living and Relating	75
Business: Organizations and Customers	75
Business, Politics & Globalization: The Core Perceptions & Behavioral Ramifications	77

Conclusions: Why a Paradigm Shift is Needed   80
Our Crisis and Possibility   81
The Matristic Culture as a Manner of Living and Relating   82

REFERENCES   87

GLOSSARY OF TERMS   89

INDEX   93

ABOUT THE AUTHOR   97

PREFACE

The content of this book is to explain experientially and conceptually the theories of living systems. These theories are the work of Chilean Biologists, Dr Humberto Maturana and Dr Francesco Varela. The theories have been expanded on in several areas, particularly human relations, emotions, perception, the construction of worlds, the constitution of cultures and the connection between the boundary and structural coupling – how two living systems change and adapt together in congruence with the medium.

These expansions are of course heavily influenced by Dr Maturana's work. I do not claim in any way that the conceptual and experiential understandings of the construction of worlds and cultures are the work of Maturana and Varela, rather, they are expansions.

This experiential and conceptual understanding arose from many years of reflecting, studying and applying these theories to my own life. I began to see how the theories were reflecting daily life experiences that I had been unaware of before. I had been culturally blind to this way of being, living and relating as a biological living system. I asked myself why, and concluded that the experiential blindness was a result of my cultural conditioning.

Never before has there been an explanatory paradigm or worldview that explains what it is to be a human being, a biological living system, and the consequences that has for our daily living in human relations, our culture and finally the human species, homo sapiens, sapiens.

Jane Cull, August, 2000.

ACKNOWLEDGEMENTS

I would personally like to thank Prof. Fernando Gonzalez from the Universidad Autonoma de Sinaloa in Mexico. Some of the material in this book was derived from a seminar that we co-presented on the Biology of Cognition and the Biology of Love, held in Boston, Massachusettes in April 1998. My thanks also to Manuel Manga and Rick Karash for organizing and holding this seminar – your faith and trust in this work, and us, is appreciated and not forgotten. My thanks also to Hugo Urrestarazu who helped with the finer details of the structure and organization of systems.

To all my friends who have supported and encouraged me throughout the years – Ian David McPherson, Meran Glufski, Zoe Harvey, Colleen de Winton, Louise Anderson, Heidi de Silva, Paulo Garrido, Fernando Gonzalez, Manuel Manga, Steven Hoath, Tim Sullivan, Thomas Hughes,

Pille Bunnell, Andre Fiedeldey, Alex Riegler, Kent Palmer, Ouafa Rian, and my parents. To the African men in my life who influenced my understanding of the relational world – Rahim, Talla Ndiaye and Pasipanodya Jani. My thanks also to Judy Sunnex who did all the fine detail editing of this revised version (2013). And finally, Dr Humberto Maturana for his friendship and assistance over the years in coming to grips with some of the deeper understandings of his work.

CHAPTER 1

THE ONTOLOGY OF THE OBSERVER

What does Ontology mean in the context of our daily lives? Why is it important? Ontology is a term that is used to describe or explain our existence and the world around us. The ontological diagram of the observer, see page 11, shows how we live two very different explanatory, experiential worlds – objectivity-without-parenthesis and objectivity-in-parenthesis. Thus, understanding our ontology, being and existence has consequences for our way of being, living and relating.

These terms objectivity-without-parenthesis and objectivity-in-parenthesis refer to the existence of objects, how objects are brought forth or constituted by the observer. It is a question of how the observer does observing, either as a biological process (objectivity-in-parenthesis) or as a result of properties that the observer has (objectivity-without-parenthesis), i.e., mind, energy, consciousness, god, soul, self, etc., properties that are assumed to be inside the body determining what we do. Thus these explanatory paths are different kinds of explanations of what it is to be human, how we are and what we do as human beings, as biological living systems.

The question of observing and the observer are also questions about reality, how we live and experience the 'world'. For example, in objectivity-without-parenthesis, or brackets, the existence of objects is lived as being independent from what the observer does, constituted by the properties mentioned above. Objectivity-in-parenthesis, or brackets, on the other hand, is the awareness of what we do as observers in language, we bring forth objects in distinctions in language. Therefore, the existence of objects depends on what we, as observers do.

We move between these two explanatory, experiential worlds, objectivity-without-parenthesis and objectivity-in-parenthesis according to our emotions or emotioning (the doing of emotions). It is in this background of emotioning that we bring forth realities, experiences as distinctions, as domains of explanations, descriptions etc. Usually however, we don't see the emotions because we think that we are rational beings, i.e., perception has to do with a property that the observer has, the mind, a pre-existing entity that is in the head and determines or controls perception and behavior.

In the explanatory path of objectivity-without-parenthesis, we live as if objects are in themselves, pre-existing and independent of what we do as observers in language. When we perceive in this way our thinking and relating changes. We think and relate as if everything is real and true in itself, there is an ultimate reality that we are all supposed to live and agree upon. This reality validates our explanations and experiences and we argue about our experiences of reality – who is right and who is wrong.

In objectivity-without-parenthesis we live and explain our experiences of being and existence as a transcendental ontology. That is, being and existence, reality, etc., is independent of what the observer does.

In objectivity-in-parenthesis we become aware that everything that we live and experience is brought forth by us in distinctions in language. In other words, we are constituting/constructing a world and our experiences in human relations, in explanations.

Every explanation is a reality. Thus we live a multitude of realities, a multiverse. And as explanations have to do with human relations, every explanation that we bring forth is valid because we are constituting/constructing the experience through the explanation.

As the world appears in what we do, in what we bring forth in language, we begin to see that the world is relational, because we are constituting or constructing it together in human relations. This is difficult to see however, as we mostly live and relate in objectivity-without-parenthesis and remain blind to what we are doing in language.

In objectivity-in-parenthesis however, we live and explain our experiences of being and existence as a constitutive ontology, the awareness of what we are doing in language, how we bring forth, construct or constitute a world of objects in language. Reality is thus constituted by us, by what we do. Reality is a social construction brought forth by observers in human interactions, or relations.

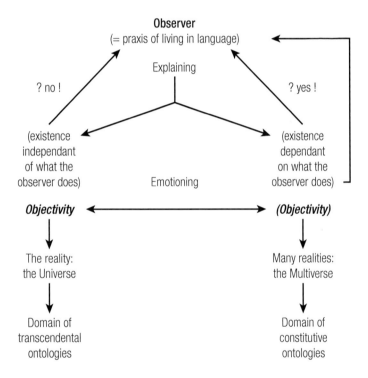

To begin the exploration of the diagram shown above in more detail, we will start at the top, the observer, in the doing (praxis) of living in language, bringing forth explanations and experiences as realities. Living, experiencing also happens to us. We are in the experience of living. We don't have to put any effort into it; we just do it naturally.

What is experience? When we distinguish, bring forth, what is happening for us. That is experience. We live the experiences of living: of being happy, of being bored, of being tired, or being in despair. We are observers bringing forth what is happening for us in our living.

How come we can do whatever we do as observers though? How come we can do observing? In asking these questions we are looking for an answer, an explanation that explains how come the observer does whatever he or she does.

In the explanatory path of objectivity-without-parenthesis the explanation would be; the observer does observing through pre-existing properties or endowments that the observer has, e.g., mind, consciousness, energy or self. This kind of explanation doesn't really explain how come the observer can do observing, because these properties are assumed to pre-exist independently of what the observer does. Mind, consciousness, energy, or self, as properties or endowments, are there in themselves, independent of what the observer does. Thus the explanation is rejected. This is what – ? no! – means in the diagram.

In the explanatory path of objectivity-in-parenthesis however, we are looking for a process, a generative mechanism that explains the experience of observing. The term 'generative mechanism' refers to the experience or phenomena to be explained.

Maturana proposes that there are two conditions or aspects that need to be satisfied for an explanation to happen, otherwise we will not have an explanation. Maturana distinguished these two conditions or aspects as A and B.

A is what is distinguished as the formal part of the explanation that gives rise to a generative mechanism that generates the experience or phenomena to be explained. For example, a little girl asks her mother, how come I was born? This is the experience or phenomena to be explained. How come I was born? What is the generative mechanism that the mother proposes to the little girl? Well, the stork brought you from Paris. This is an explanation.

The formal part of this explanation is in the form of a generative mechanism or process that as a result of its operation we have the experience that we want to explain. This is quite different to descriptions. Descriptions describe experiences, they don't explain. For example, the mother would describe the scene to the little girl on how come she was born in this way. "You were born in this beautiful gooseberry patch. There were lots of flowers there and I remember seeing butterflies hovering just near you."

There is a second condition, B, however. This condition is very important, because without this condition there is no such thing as an explanation. This condition is that the observer accepts an explanation according to his or her criteria. Thus, explanations are not in themselves.

Explanations become explanations when an observer accepts the answer or explanation according to his or her criteria for what constitutes an explanation. And this of course depends on how we as observers listen.

This is indeed what we do in daily life when we do or don't accept explanations according to our criteria. If we accept an explanation we are tranquilised, we are satisfied. If we are not, we go on asking questions. So explanations have to do with what we do as observers. It is difficult to see the doing though when we think and interact in the path of objectivity-without-parenthesis, because we think that explanations are 'right' in themselves. This of course obscures the criteria of the observer in which the observer rejects or accepts an explanation.

Everything is said by an observer. (1) It is the observer that constitutes explanations in formulating processes, generative mechanisms to explain experiences. Thus experience is constituted by what the observer does. We explain experience from or with experience. This is what we do. We bring forth or constitute our experiences through explanations in language.

No thing is in itself, not even the observer. The observer does not exist in the body doing observations. Things, objects, entities including the observer are distinguished, brought forth in language.

Objectivity-in-parenthesis reminds us of what we do, the actions or behaviors of distinguishing, how we bring forth the existence of objects in distinctions in language. The parenthesis refers to what we do, how the constitution, existence of objects arises.

For example, imagine we are typing something, an email perhaps. Everything that is in the email has been brought forth in actions, behaviors of distinguishing. Words, thoughts, are actions of distinguishing. Where is the distinguishing taking place? The distinctions are occurring or happening in the behaviors of typing, relational behaviors that are dynamics between us, the keyboard and the screen. We never do any behavior on our own. All behaviors arise in the relation, the interaction.

In these actions or behaviors of distinguishing, we are constructing/constituting our experiences. We live these experiences through what we construct and explain and are lived as a reality. Thus, we live a multiverse or many realities and, as we have constructed a particular reality, it is valid in the domain in which it exists, i.e., in our explanations.

Explanations constitute different domains of co-existence of living and doing together, behavior. We accept and reject with responsibility explanations according to our criteria, our liking and disliking with the awareness of our criteria for accepting or not accepting. This is the awareness that explanations are not in themselves, of being right or wrong, rather we accept or reject according to our criteria. In this way, explanations, explaining experiences has to do with human relations – how we are relating.

Through understanding how we are constructing or constituting our experiences, realities, explanations in behaviors of distinguishing in language, we can see that the observer, or more precisely observing, arises as a part of this process. That is, how we live in language, the bringing forth of objects and entities in actions of distinguishing is a process of observing.

Observing is thus a process that we do in our living. It is a never-ending recursive, circular dynamic that appears in our many actions or behaviors of distinguishing either in thinking or in human relations. The arrows on the right hand side of the diagram from explaining to existence and back again to the observer, reflect this circular recursive dynamic.

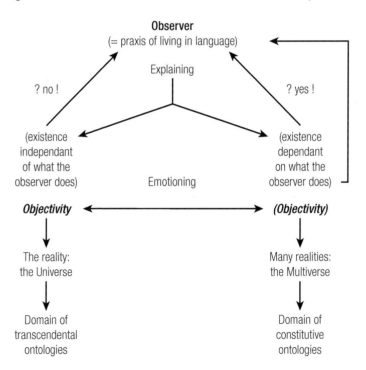

Scientific Explanations

Scientific explanations are not that dissimilar to everyday explanations. Instead of two conditions, there are four:

a) First is the experience to be explained which is what we distinguish as a phenomena – the phenomena or experience to be explained.
b) Then an explanatory proposition or hypothesis is proposed to a body of observers that generates the experience to be explained.
c) Other experiences that were distinguished in b), but may not have been considered as part of the initial explanatory proposition are now being considered as conditions for an observation by a body of observers.
d) Observing the experiences or phenomena that were distinguished in b).

Only when these four conditions are satisfied will there be a scientific explanation. For example, let's say there are four observers doing physics together and they are generating the experience of gravity as the phenomena to be explained (a). In this process they bring forth other conditions or experiences like force and attraction between planetary objects or entities (b). As part of describing these experiences, other experiences may be distinguished such as electromagnetic fields and this becomes part of the conditions of the explanatory proposition (c). The conditions distinguished in (c) become part of (b) as experiences for observation.

This kind of explanation has to be accepted by the criterion of validation as proposed by the physicists. What that means is the scientific explanation is only accepted within the domain of physics because the observers in that domain constituted it as such.

Emotioning and the Two Explanatory Paths

To complete the exploration of the ontological diagram, there are two arrows coming off 'explaining' with the word 'emotioning' underneath.

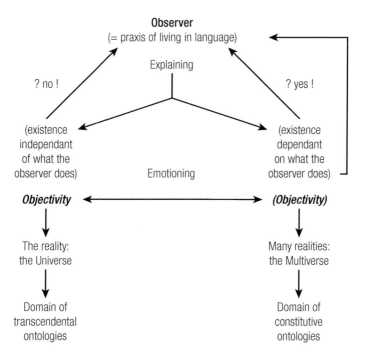

This term, emotioning, refers to our biological dispositions, our moods, as Maturana defined it, that alter our manner of distinguishing, thinking and relating in language, as well as our general behaviors. It is an understanding of how we move or flow between the two explanatory paths according to what is happening in our emotioning.

Emotioning is the doing of emotions. Emotioning appears in what we do as different kinds of behavior, different kinds of classes or actions of behavior. The doing of emotions, emotioning is important to see, because we are constituting and constructing worlds, experiences through our explanations in this background.

For example, what do we do when we claim objectivity? What are the emotions in which we make these kinds of claims? Claims made in objectivity appear in the emotions or emotioning of control, the desire for certainty in which anger and frustration generally appear. We claim a privileged access to an external and pre-existing reality, e.g., I know how things really are, you don't know. We become righteous as an authority on how reality is. Reality from the emotioning of control is experienced and lived as transcendental, reality exists independently from what the observer does. This manner of distinguishing in this kind of emotioning has consequences in human relations because when we claim to be the authority on a transcendental reality we deny another's experience or explanation. This experience of negation is lived as a negation of self (identity) or, in some cases a negation of 'beingness', how they are is transcendentally 'wrong'.

In objectivity-without-parenthesis we live as if experiences and explanations are in themselves, we are those explanations and experiences. As such they become associated with identity. For example, if someone tells us that we are wrong, we live and think that there is something intrinsically wrong with us. "They don't like me, what I say is not good enough." We can live this as a fundamental rejection of who we are.

Disagreements and arguments also appear in the desire for control and certainty, e.g., who or what is right or wrong. Certainty is backed up with proof of what is 'real in itself'. Proof validates our reasons, explanations. "How can you disagree with the data, the proof. I mean, look, get real, there it is, how can you disagree?" "Yes, but if you see it in this way, then you will understand what I'm saying." "No, you're wrong, because the data says something else. The data is right and you are wrong. You are being subjective and not seeing reality for what it is."

Religious wars are also justified through what is written in the Bible or the Koran, or by what religious leaders say, as proof of 'reality'. If it's in the Bible or the Koran, its true, real and intrinsically right in itself. You must live and act in this way because the Bible or Koran says so. "This is reality and what you are living is a sin against God. If you don't live and behave in this way, God will punish you." Notice though how God as a distinction is used as a justification to suppress or dominate others.

In objectivity-without-parenthesis knowledge is lived as relations of mutual denial. We deny the other if they don't think like us. "This is 'the way' to think and the rest of you are wrong, dumb, stupid if you think differently." The only thing left is submission, denial.

So objectivity-without-parenthesis is about fundamental truths, beliefs, perceptions about what is 'there' in reality. Scientific and other kinds of discoveries are claimed in this way as well. For example, scientists have discovered a new mathematical law of nature – the law is there in itself waiting to be discovered. Sounds a bit funny when you think about it as laws are brought forth by us in language thus they are not 'out there' waiting for us to 'discover' them.

In objectivity-in-parenthesis we live or constitute a very different world, a world that is dynamic, fluid and ever-changing. The 'world' as is stated here, reflects the many behaviors, distinctions that we do, many realities. We live and relate in love, trust and acceptance; whatever happens is whatever happens. This is what we distinguish as the 'flow' or going with the 'flow'. Our perceptual awareness is expanded as we bring forth many possibilities or ways of seeing situations. We don't try to control what is going on, our own behavior or someone else's. The desire for certainty and control does not appear.

To live and relate in the flow is to be congruent with the many perceptual behavioral changes that are happening in thinking and human relations. And these changes happen or appear as a result of what is happening in the body, our biological structure. As soon as we begin to suppress or control our thinking or behavior, we control or suppress our biological activity that is happening in the moment. And this is how we constitute the explanatory path of objectivity-without-parenthesis.

In objectivity-in-parenthesis we become more aware of our bodies, biology, and also what we are doing in language and behavior generally. We find ourselves reflecting on our experiences, in what we are saying or thinking, distinguishing, or how we are relating with one another and, in this operation or process of reflecting we become aware that we treating our experiences as objects, we are observing what we are distinguishing, i.e., experiences and explanations.

When we do reflecting and observing in this manner we become curious. The desire to learn and explore becomes primary instead of the usual desire to control and manipulate the relation through rationality and competition, i.e., proving and justifying what is 'right' or 'true'.

Everything that we do and live in objectivity-in-parenthesis appears as an expansion of behavior in the background of emotionings, love, trust and acceptance. Love as a bodily disposition expands the plasticity of the nervous system resulting in an expansion in behavior.

How is love defined in this way? Love or loving is to live in legitimacy, the legitimacy of another in co-existence with oneself as Dr. Maturana frequently distinguishes. To live in legitimacy, in loving relations, is to validate and accept not only what is happening in any given moment for ourselves but also for others. We can live in compassion and understanding with the awareness that all of our behavior is appearing as a result of biological or bodily activity.

When we listen and relate in loving relations, we are validating the world of the other, what they are constructing as their experiences through their explanations. We do these kinds of relations in friendship. In friendship we live in acceptance and legitimacy, equality and mutual respect. As soon as we start to control or demand, friendship disappears and another kind of relation appears – denial. Suddenly we are back living and relating in objectivity-without-parenthesis once more.

All of these reflections pertain to daily life as we move between two experiential and explanatory paths or worlds according to the flow of our emotioning. An analysis of what shows up in our emotioning and the kinds of worlds that we constitute as a result, are shown in Chapter 4, The Construction of Worlds and Chapter 5, The Constitution of Cultures.

However, this awareness of what we do is important if we are concerned or serious about transforming our manner of living, a manner of living that has been conserved over many, many generations in the explanatory path of objectivity-without-parenthesis. For in this explanatory path we have lost awareness of our biological roots and our mutual co-existence with the environment. Perhaps it is time to embrace and reawaken our systemic awareness and understanding of life from the explanatory path of objectivity-in-parenthesis, the powerful explanations brought forth by Maturana and Varela, the theories of living systems.

CHAPTER 2

THE OPERATION OF LIVING SYSTEMS

Defining Systems: Simple and Composite Unities

What is a system? Or rather how is a system defined? In daily life we can distinguish systems or unities in one or two ways, either as a simple or a composite unity.

The systems that we distinguish as simple unities are those that appear with properties. Simple unities do not have components. We distinguish the unity or system for example, a rock, according to its properties, i.e., its color, its shape, its size in terms of dimensions. A business card is another example that does not have components, but we refer to its properties (the texture, board or plastic, the color, words or text) in describing the system or unity.

The other kind of system or entity that we distinguish in daily life is a composite unity and we see this when we distinguish components and relations, for example, a pen. If we take a pen apart we can see that it has components and that these components are related with each other that constitute the object, pen.

It is these components and their relations that constitute how the composite unity is made or put together and how it operates as a whole, a total unity, a system. Without these components and their relations, the pen would not work and we would not be able to perform the operations of writing with the composite unity or system that we call pen.

From this example, we can see that two domains of existence have been distinguished. One is how the pen is made or put together as a composite system or unity. The other is the environment or medium, where the pen exists in the domain of writing, in the domain of interactions, with an observer moving the tip of the pen over the surface of a medium, i.e., paper.

The Characterization of Living Systems: Autopoiesis

The term autopoiesis refers to the characterization of a living system and defines the organization, i.e., the components and their relations. These components and their relations are constituted by the processes of autopoiesis – molecular and cellular self-production, how a living systems organization and structure arises.

The processes of autopoiesis also specify the network of the system, the extension of how far the network will go.

This is the boundary of a living system. The boundary that is specified by autopoietic processes enables the operational closure of living systems.

We can see the phenomena of autopoiesis in daily life when we see the cells of the sperm and ovum couple, triggering changes in their structure such that new cells and molecules are generated or produced. The cellular and molecular network expands and we begin to see the outline or shape of a human fetus as it grows along with the components and their relations that constitutes its structure.

This growth, otherwise known as the gestation period is a direct result of autopoietic processes. However, the autopoietic processes do not just stop there. Our hair and nails grow throughout our entire life cycle. The healing of a wound as the cells coalesce and bind around the opening also reflects these processes.

As a result of these autopoietic processes as mentioned earlier, a boundary or surface is formed. This is the extension and closure of the network.

Even though we are operationally closed as a result of the boundary or surface that is constituted through the processes of autopoiesis, we are open to matter in the environment. For example, we breathe and exhale air through our mouths and noses. We also drink and eat through our mouths. As we do these behaviors, the matter becomes part of the structure, the circular dynamics.

Matter however, does not determine what happens in the structure. Matter influences or triggers structural changes in the network. But it is the structural changes that determine what happens in the network, not the matter itself. This is reflected when we breathe polluted air for example.

For some of us, this is not a problem. But for others like asthmatics, it is a very big problem as they may experience difficulties breathing. So here we can see that the air particles or matter are not determining what is happening in our structures. Rather matter or air particles trigger changes in the structural dynamics among the components and their relations.

The Organization & Structure of Systems

The term structure refers to a composite unity/system that we, as observers, distinguish in daily life. Structure is the components and their relations. The term organization on the other hand is a bit more subtle, and refers to the relations between the components of the system/unity that gives rise to its class identity, a particular class. Organization and identity go together and it is the organization that is conserved that gives rise to a particular identity. It is important to recognize however, that it is us, as observers, that distinguish the identity of systems. Identity is not there in itself. We bring forth identity in language.

To see structure and organization more clearly, lets take a simple example from daily life, a chair. We distinguish or recognize a chair as a particular identity or system due to its organization, how the components and the relations (back, seat, legs) are put together such that we distinguish this kind of system as a chair. It's organization is conserved unless the chair is cut in half for example in which case the chair as a particular identity does not exist anymore and becomes something else. Whilst the structure, the components and their relations of the chair can be made from many materials, i.e., glass, wood, plastic and metal, these do not affect the identity, organization of the chair.

In the context of social systems, structure and organization can be distinguished in many ways. There are many types of social systems that arise in daily life and can be distinguished as different kinds of social organizations, i.e., a church, a family, a business, an army, etc. Social organizations can be formed either spontaneously or around a common purpose or vision. There will be particular patterns of behavior that are conserved among the individuals that constitute the structure of the organization. For example, the structure could be hierarchical. A hierarchical structure is defined by hierarchical relations, how individuals interact and relate with one another (particular patterns of behavior that are modulated by particular kinds of emotions, fear, mistrust and control for example). Even when a human social system emerges spontaneously through the interplay of languaging interactions (human relations, conversations) between individuals, its structure may be extremely variable in time whereas its organization, if conserved through time, may tend to be very stable (this is the case of family structures across long pre-historical periods).

How Living Systems Operate: Structural Determinism

Structural determinism refers to the operation of living systems. All living systems operate according to their structure, to how the structure is made or built. As such, living systems do what they do as a result of what is happening in the structure, to the changes that are taking place among the components and their relations.

As living systems, we do whatever we do according to our structure, to what is happening in it. Our structure is a dynamic network of interconnected activity that is taking place between the components and their relations. It is always changing as a result of what is happening among the components and their relations.

All living systems have a boundary, a surface or skin. The boundary, surface or skin, is the closure of a living system. We operate as closed living systems. Thus, the medium or environment does not inform or determine what happens in the structure. The medium or environment can only trigger changes in the structure.

Let us see where structural determinism shows up in daily life. We all know that when we are not feeling well, that there is something wrong with our structure, our biology. So we go to the Doctor, someone who understands how the components and their relations, the structure works. We ask him or her to make an assessment of what is happening in our biology to explain why we are not feeling well.

We may also notice when we are unwell, changes in our emotionings and perceptions. We may feel grumpy, irritable or even depressed. Life does not look particularly rosy when we are not feeling well. These changes in our emotionings and perceptions reflect structural determinism, i.e., changes taking place in our structures.

We can see this as well with other systems that we interact with, like computers for example, especially when they don't work. We turn it on and nothing comes up on the screen, or we are in the middle of typing something and the whole system crashes. As a result we lose all of our work.

What is happening there? Are we determining what happens in the structure when we try to get it to work? Are we giving instructions to the computer? Of course the answer is no. We are not determining what happens in the structure of the computer. We are only triggering. The structure of the computer does what it does according to how it is made or built.

Now our awareness of this situation is reflected when we take the computer to a technician or computer shop to have it fixed. The technician is not looking at our hands or fingers for the problem. He or she is looking in the structure, at how it is made or put together. He or she looks at the components and their relations to see where the problem is.

At the same time, we don't determine what is happening in the medium either, because the medium also operates according to its structure. The medium is also a structure determined system. For example, we can see this when we interact with plants in the garden. We may spread fertilizer to assist the plant's growth, and the plant dies. In other cases, the plant may grow voraciously. What is apparent though, is that we are not determining what is happening in the plant through what we do, i.e., spreading fertilizer. The plant does what it does according to its structure.

The consequences of structural determinism in human relations are very important. It is an experiential awareness that in the moment we cannot distinguish between perception and illusion. Nor can we determine or control what other people do, their behavior. We can influence, yes, but not determine.

As a result of structural determinism we live in perceptual and behavioral uncertainty. This is very apparent in daily life and is sometimes a source of confusion and frustration in human relations. This confusion and frustration disappears with the understanding and awareness of structural determinism. These consequences are explored in more detail in Chapter 4, The Construction of Worlds.

The Nervous System

The nervous system operates also according to its components and their relations, its structure. The structure consists of a vast network of interconnected activity that takes place between the neuronal components or elements. It is these components or elements and their relations that constitute the neuronal network that is the nervous system.

The relations of activity among the components (neural elements) generate changes in the neuronal network. These changes then lead to further changes in the relations of activity between the components and continue in never-ending recursive or circular dynamics.

These recursive dynamics appear in our behaviors as repetitions or habits, patterns of repetitive or conserved behavior. And as we know from daily life, we do many patterns of behavior in human relations and also varying environments or mediums. We are able to do these many and varying patterns of behavior as a result of the plasticity of the nervous system.

As the nervous system is operationally closed, it does not operate with representations of the environment. There are no words or pictures floating around in the brain and nervous system. Thus it is not the environment or medium that specifies or determines what happens in the nervous system. Rather, the nervous system does what it does according to what is happening in the relations of activity taking place in the neuronal network.

The understanding of how the environment or medium influences or triggers changes in the nervous system will be explored in Chapter 3, Living Systems and The Environment.

Function & Purpose

Generally, in our culture, we speak about the function or purpose of systems. These distinctions are misleading because they obscure the operation of a system.

Systems operate according to how their structures are made or built. As such they do what they do as a result of their structures and not because of a purpose. The distinction of purpose belongs in the domain of the observer. The distinction of function though is similar, although it doesn't take into consideration the relations between the components. Function just generally refers to components; thus the relations are obscured when this particular distinction is used.

CHAPTER 3

LIVING SYSTEMS AND THE ENVIRONMENT

Co-existence: Structural Congruence with the Medium

Every living system that we see in daily life is distinguished in two domains of existence, the domain of the biology and the domain of interactions (behaviors) with the medium or environment. We see the biological dynamics or changes that any living system undergoes as behaviors that appear in the domain of interactions.

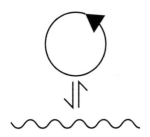

(From The Tree of Knowledge, Humberto R Maturana & Francisco J Varela, Shambhala Publications, Boston, Massachusetts, 1992)

This simple diagram shows the life of a living system as a continual ongoing relational dynamic between the two domains of existence. The arrows in the diagram, one up, one down, reflect the congruent or complimentary relational dynamics of a living system encountering or interacting with the medium. The arrows do not mean interchange or exchange, nor input or output. The circle with the arrow reflects how a living system operates, as a closed structure determined system.

These two domains, the operation of a living system and its interactions or behaviors with the medium, do not intersect. One does not cause the other. Instead there is a generative relation between them.

This generative relation refers to how a closed living system changes and adapts congruently in interactions with the medium or environment. Both mutually transform or adapt in interactions. Inputs, outputs, exchanges, cause and effect, obscures this dynamic of mutual change and adaptation, transformation between living systems and the environment.

This process of mutual transformation, change and adaptation occurs through the structural coupling of the surface of the living system, the boundary; and the surface, boundary, of the medium. This coupling takes place in recurrent, repetitive interactions. This is how the medium (environment) influences or triggers changes in the structure of a living system. The biological patterns are conserved through the process of structural coupling (change and adaptation) in recurrent, repetitive interactions with the medium.

Biological and behavioral change is a constant for living systems and the medium as both co-exist in an ongoing structural drift of congruent structural changes (change and adaptation) in recurrent, repetitive interactions. And it is at the boundary, the skin, where we see the process of change and adaptation taking place as changes in behavior in the interaction with the surface of the medium.

In the picture below, we can see this dynamic when we run along a beach. When we run we are structurally coupled with the surface of the medium. The boundary or surface of the living system, the feet in this case, are coupled with the surface of the medium (the sand) in ongoing recurrent interactions. In the structural coupling of surfaces we can see changes in the behavioral patterns as a reflection of change and adaptation with the medium, as footprints left in the sand.

Usually we don't see the generative relation of mutual change and adaptation between the two domains of existence

in daily life, as we take it for granted. It is something that we do naturally in our systemic interconnectedness with the medium.

In order for any living system to be alive, two things must happen simultaneously: the conservation of the system's organization (biology) and the coupling between the system and the medium. When a living system loses either its coupling, its congruence with the medium, or its organization, the living system will die.

All living systems do not live in isolation from their mediums/environments, they live in the relation between the bodyhood and the environment/medium. This reflects the natural interdependency of co-existence between all living systems and the mediums/environments in which they exist. If the environment is heavily polluted with toxins and chemicals, the chances are that the living systems will die, i.e., there will be a loss of congruence with the medium/environment in which they co-exist. All living systems do not live in isolation from their mediums/environments.

Interactions: Triggering, the Bodyhood & Behavior

When a living system encounters the medium in an interaction, the medium only triggers structural changes in the system. The medium does not instruct, give or send information to the system. Whatever happens for the system as a result of the triggering depends on what is happening in its structure and not by the medium.

The term 'encountering' is also relevant to triggering as the medium or environment does not pre-exist. The medium or environment appears or exists in the encounter between a living system and the medium or environment moment, by moment, by moment. Both the living system and the medium or environment change and adapt together congruently. Thus one does not pre-exist the other, but shows up or appears in the moment of encountering.

Human relations is often characterized as communication, i.e., we are sending and receiving information, we are determining what someone hears. This distinction, communication and what it implies, that we are determining what someone hears through the sending and receiving of information, is misleading however. It becomes apparent that this is not the case in many examples of daily life. For example if we are speaking with a person in English and they only speak another language, they do not understand what we are saying. If we are speaking with someone who is hard of hearing or deaf, they will either only partially understand or not understand at all. In these examples, it becomes obvious that we are not determining what they hear. People will hear what they hear according to what is happening in their biology, their structure in any given moment, hence we are only triggering, not determining in interactions, human relations.

Every interaction that we participate in, every behavior that we do in the relation or interaction, takes place in an ongoing present. We do not do past or future behaviors. No living system does. All behaviors, interactions, take place in an ongoing present. The medium also does not exist in the past or the future, but arises in the encounter between itself and a living system in an ongoing everchanging present.

We live in an everchanging and ongoing present. This is contrary to what we are culturally led to believe, that we live an independent and pre-existing past and future. The distinctions of past and future arise in the domain of the observer in language to explain the coherences of experiences. As such the past and future do not exist as entities in themselves independent of what the observer does in language.

The behavior of living systems appears in the relation between the bodyhood (the body) and the medium in an interaction. Thus, behavior is relational. Behavior is never something that a living system does alone. Let's take a look at an analogy to see this dynamic of behavior.

We pick up a lighter to light something and we trigger a particular part or component of the lighter to ignite the flame. It is apparent that we are only triggering the structure of the lighter, because when the lighter is low on gas, a flame does not appear.

Now when we trigger the particular part or component of the lighter, where does the flame appear? Does the flame appear in the lighter? No. It appears in the behavioral domain between the lighter and the air, in the spatial relation.

Normally we think that we do behaviors alone. This is not the case however. For example, if we hang from a branch in a tree and do the behaviors of walking are we walking? No. What happens when we climb down from the tree to the ground? Do we do the behaviors of walking in relation to the ground? Yes.

Walking as patterns of behavior appear in the encounter with the medium or ground in recurrent interactions. There is no separation between the bodyhood and the medium as both surfaces or boundaries are structurally coupled in a process of mutual transformation in recurrent interactions. And we see this transformation as changes in the patterns of behavior that we do in the relation or interaction with the medium.

This systemic and circular dynamic between our bodyhoods and the medium is not seen in objectivity-without-parenthesis. This is because we see behavior as being separate to the body and also the medium. Behavior is also perceived as being fixed and unchanging and follows a linear line according to time. However, it is not time that determines the flow of behavior, but rather the changes in the biology (structure) that are triggered as a result of recurrent interactions with the medium.

Recurrent Interactions and Learning

The next situation we will explore is recurrent interactions. This is where learning takes place. The foundation of recurrent interactions is of course, structural coupling, so when we interact recurrently with the medium, we begin to repeat actions, doings or behaviors. In other words, learning is a doing and takes place as repetitions of behavior that if repeated often enough become conserved such that the behaviors become automatic and natural. Examples from daily life are learning to drive and steer a car, learning to ride a bike, learning all sorts of sports and activities, learning another language, learning to use a computer, learning tasks, learning subjects at school and university etc.

The same situation applies to human relations. We will repeat various behavioral patterns in the work place, socially or in intimacy. Normally however, we do not see these behavioral patterns because our focus is not on what is taking place in the interaction. Rather, our focus is on the person that we are interacting with and as a result we are not seeing what is going on in recurrent interactions, i.e., how we are influencing each other structure's in mutual change and adaptation, establishing biological and behavioral patterns in recurrent interactions.

Change in behavioral patterns takes place around what is conserved. Thus, to change or influence the behaviors that we do in our interactions, we need only be aware of what we are doing through reflecting. When we reflect on what we are doing in our interactions, we accept what is happening. Thus, the behavior changes automatically and if the behavior is repeated, that pattern of behavior will become conserved biologically as a habit or routine.

Understanding Behavioral Change

Changes in our behavioral patterns occur in various ways. The first is naturally as a result of biological processes of activity triggered either by what is happening in the biology or as a result of interactions, i.e., change and adaptation with the medium. The other is through what we do in language. In language we reflect in our thinking or in a conversation.

As we reflect, we open a space where we let go of certainty (control) and move into the dispositions of trust, acceptance and curiosity. As we do that, we can choose behaviors according to our preferences – what we like or don't like or what our bodies do or don't like so to speak, what does or does not feel comfortable.

Change cannot happen through demands or force as fear or control inhibits the plasticity of the nervous system. Fear and control limits or inhibits the range of possible behavioral patterns that we can do, i.e., reflecting. In fact, experientially I have found it very difficult to even reflect in the emotioning of fear and control.

When we accept or validate patterns of behavior that we do in any given moment, change occurs naturally. This understanding I feel is very important. It is an awareness of how we are as structure determined living systems and as such we cannot be doing anything other than what is happening in the moment. What is happening is what is happening. In the moment we cannot change what is happening. Change always occurs after the moment, a posteriori. In that sense, we are living a series of changes in our behavioral patterns. Change is constant. If however, we resist what is happening in the moment, the pattern will persist and not change. Through the emotioning of resistance, fear and control, the plasticity of the nervous system is inhibited. Thus changes in the biology and behavioral patterns are inhibited also.

Understanding how our patterns of behavior are conserved biologically in recurrent interactions enables us to understand our history, how our history as a manner of living and relating is conserved. In daily life we can see a manner of living anywhere where there is a history of interactions involved, e.g., two or more people living together, families, couples, businesses, community organizations, schools and universities, etc.

Throughout this history every individual is influencing another's evolution. We are co-evolving together in recurrent interactions in which a history is established and conserved in biological and behavioral patterns. Our co-evolution, how we are co-evolving, is reflected in the conserved patterns that constitute our manner of living.

From this we can be aware of how we are influencing one another in interactions and what consequences that may have for our evolution as a whole, as a species.

Language

Language or languaging (language is a doing hence languaging), is also something that we do in our interactions and pertains to the two domains of existence, the bodyhood and the medium.

Normally we think of language as symbols that exist inside our brains as representations of an external reality, representations that the brain has received and processed in terms of information. This is the classical objective view. Language is not taking place in the head or the brain as we normally think though. Language as a manner of existence takes place in the domain of behaviors, in what we do. Every distinction that we bring forth as observers arises in language or languaging, the doing of language in our manner of living. Normally we think that thinking is taking place in the brain. However, if we pay attention to what we do, we can see that thinking arises in distinctions in language and is thus a relational, doing, behavioral phenomena. Similarly when we are conversing, we are conversing in distinctions in language that we bring forth as observers. We explain, share, describe, bring forth objects in distinctions in language in conversations in human relations. As such languaging does not appear in the bodyhood or structure, but appears in the doing, in the relation, in the domain of behaviors, in the recursive interactions between the bodyhood and the medium as a spontaneous flow of doings or behaviors in recurrent interactions. Indeed, languaging as a recursive process appears as consensual co-ordinations of co-ordinations of behavior.

Some examples from daily life of consensual co-ordinations of co-ordinations of behavior are: a group of musicians playing together; African tribes communicating or co-ordinating with each other through the sounds or natural rhythms of the drums; how we naturally co-ordinate our behaviors together congruently in dancing; and even in those moments of pure intimacy where we understand each other without words. In more simple circumstances we may ask someone to get something for us. We ask them to get an object, whatever it may be, they turn and find the object and bring it to us. Those actions are consensual co-ordinations of co-ordinations of actions.

This consensuality of co-ordinations is not so easy to see in daily life because we think that we are sending or receiving information or consensuality has to do with how we are communicating with one another, whether the information was received properly or not. However, if we see our interactions in another way, as co-ordinations of co-ordinations of behavior, not of words tooing and froing between us, then we can see the dynamics differently, how we co-ordinate our actions together in language, or to be more precise languaging.

The Constitution of Human Beings

This is the final section of the two domains of existence, the bodyhood and the medium. Where do we exist as human beings? Do we pre-exist as human beings? Does being human have to do with our bodies or bodyhood?

We are born with a particular bodyhood, homo sapiens, sapiens. That's the classification. To be a human we have to have two domains together: a bodyhood that is homo sapiens, sapiens, and language. So it is not only language, nor the bodyhood but both.

The manner of living that we live in language as human beings is not always social. In order to live in social relations there has to be a particular emotion present. That emotion is love. Love is the only emotion that allows us to live with the other in mutual respect and acceptance.

Indeed our co-existence or the awareness of this can only be realized through the emotion of love, because the other arises as a legitimate other in co-existence with oneself. In objectivity-without-parenthesis, we are unaware of our mutal co-existence because when we relate/interact in objectivity-without-parenthesis as a manner of living and relating, we relate and interact in denial. And when we do this, the other with whom we are relating is not legitimate.

Our western culture is mostly centered in relations of denial or non-social relations. Thus our humanness is denied also as a manner of living and relating. We are not human beings in relations of denial or negation. The fact that we have difficulty with love or are searching for love and respect reflects that we are not living those kinds of relations and interactions.

High suicide rates, divorce, sexual and emotional abuse, racial and gender exploitation are just some of the other problems that we create and live as a result of denial.

We are not human beings in objectivity-without-parenthesis. We are human beings only in the awareness of our mutual co-existence, in the dynamics or interactions of trust, love, respect and acceptance.

CHAPTER 4

THE CONSTRUCTION OF WORLDS

In this chapter I am expanding on the work of Dr Maturana, particularly the ontological diagram, objectivity-without-parenthesis and objectivity-in-parenthesis, from which many of these understandings have unfolded.

In this overview of the construction of worlds, I will show how the worlds of objectivity-without-parenthesis and objectivity-in-parenthesis, are constructed as a result of our emotioning. That is, how we are constituting, constructing, these worlds in language in a background of emotioning in thinking and human relations. These reflections include:

- The process of objectification – how objectivity arises
- The consequences in human relations – perception and behavior
- The psychology of objectivity – the worldview

- The myths and contradictions of objectivity
- The relational world – objectivity-in-parenthesis
- The co-construction of the relational world – relational languaging
- The circularity of knowing & perception – relational way of living
- How we can do a paradigm shift – moving between the two worlds with ease
- The possibilities for human co-existence

World Construction Overview

The worlds of objectivity-without-parenthesis and objectivity-in-parenthesis are social constructions. The worlds appear in what we do, recursive, repetitive patterns of distinguishing in language either in thinking or in human relations, conversations. We constitute, bring forth objects, things, experiences, realities and worldviews into existence through distinctions in language. No thing pre-exists until we distinguish it in language.

The process of distinguishing in language, perception, flows according to our emotioning, i.e., our patterns of distinguishing and general behavior follow the patterns of our emotioning, cyclical processes of biological activity. This is not a one-way dynamic however as our languaging also changes our biolgical activity including our emotioning. Biolgical activity also changes when we recurrently interact in human relations, interactions with the environment, other living systems and even physical objects. As mentioned previously, this process occurs through mutual change and adaptation, structural coupling, how our bodies, our structures, mutually change and adapt together in recurrent interactions. This process was explored in detail in Chapter 3, Living Systems and the Environment.

Patterns in our emotioning can be distinguished from listening to the sound or tone of our talking in language either in thinking or in human or non-human relations. For example, in the emotioning of fear, the sound or tone of our voice changes. The sounds are shaky, high pitched etc. Similarly if we are feeling depressed, happy or excited the sound or tone of our voice will change and be different again. The sounds or tones reflect the emotioning in which we are distinguishing or relating. Notice also, how the words, distinctions in language appear in our talking and relating. For example, I feel happy, sad, depressed etc. At the same time, we also distinguishing emotioning by observing what is happening in behavior, how someone moves, their facial expressions, their gestures and mannerisms, how they hold themselves etc.

However, there is another important aspect of distinguishing – how objects (conceptual and physical) are distinguished as entities in themselves independent from what the observer does (objectivity-without-parenthesis), or as an awareness of what the observer does, i.e., objects arise or are brought forth by an observer as distinctions in language (objectivity-in-parenthesis). To show this process more clearly, we need to understand and reflect on what is going on in emotioning and languaging. For example, in the emotioning of fear (what occurs when we suppress our emotioning), our perceptions, distinctions in language change and they become lived and experienced as objects in themselves, separate and independent of the behaviors of distinguishing. These entities, objects, are commonly referred to as 'reality'. Reality as a construction becomes external to the process of perception, distinguishing in language. Experiences are lived as being true and real in themselves. We are those experiences or perceptions. The self as a distinction in language becomes objectified and is lived as an entity that resides in the body, the perceiver. The perceiver now lives a dualistic separate world, what we commonly distinguish as an internal/external world. The 'world' appears as being separate and independent from the perceiver. This is the constructed world of objectivity-without-parenthesis.

In objectivity-in-parenthesis however, the distinctions do not become objectified in the emotionings of trust, love and acceptance. Rather, the words in which objects are distinguished, both conceptual and physical, become relational, part of the flow of distinguishing in language. Reality as such is lived and experienced as a construct, just part of the flow of distinguishing, thinking or relating. As such we live in the awareness of bringing forth a world together, a shared or relational world in languaging, the doing of language. We begin to live and relate in a natural way. I feel it is important to distinguish in this moment that what I am pointing out here is the 'awareness' of what shows up in the constitutive path of objectivity-in-parenthesis. We already bring forth a world in objectivity-without-parenthesis in the relational domain irregardless of whether we are aware of that or not. What I am pointing out is the awareness of what we are doing, how we are not aware that we are bringing forth a world in language in objectivity-without-parenthesis, in other words, the world is pre-existing and we are describing, explaining it, getting to know it etc. In objectivity-in-parenthesis however, we become aware that we are constituting worlds, realities, experiences etc., in distinctions in language and we do this in thinking and in human relations, conversations.

The worldview of objectivity-without-parenthesis however, is not consistent with how we are as biological living systems. There are many contradictions and inconsistencies that will be explored later in this chapter. In the meantime however, I invite the reader to join with me on a conceptual and experiential journey of both worlds as a basis for understanding what we do, the kinds of worlds we construct in thinking and human relations in our manner of living.

The Process of Objectification: Internal/External World

Earlier I mentioned that conceptual objects become entities in themselves in the emotioning of fear. In what follows are descriptions and explanations of what occurs complete with the languaging that would be used.

The doing of distinguishing is obscured as distinctions, abstract nouns, are experienced as objects in themselves, separate and independent of the actions of distinguishing. Abstract concepts such as the self appear in this process. The self, a word or distinction in language, becomes an entity in itself in the emotioning of fear. The self becomes the perceiver, an entity doing perception. The self is now creating experience. Experience becomes separate and independent of the self, the internal world. The internal world of the self is thinking. Thinking becomes the mind of the self. The mind, another abstract concept, has become objectified and is experienced as being there in itself inside the head. So now the self is separate from the mind. The mind is responsible for what the self thinks and does. The mind controls behavior. Or, the self controls behavior through the mind.

Whatever the self thinks, perceives, the experience is lived as being there in itself as a fixed or absolute reality. In other words, the self becomes that reality and the self projects (a distinction for what occurs in the language of objectivity-without-parenthesis) experiences, what is showing up in thinking, onto other objects. Objects whether they are physical or conceptual, become what the self perceives as being external or independent from it. The self becomes afraid of this physical or conceptual object that is perceived to be external from it. Thus the self has become afraid of what he or she has distinguished whether it be a human being, an object, an animal, or the environment.

My understanding of this process arose from an experience that happened some time ago when I was lying in bed. This particular day, I was experiencing fear. I didn't want to get out of bed. I perceived my room and the environment outside of my room as being frightening. I began to reflect on this experience and realized that this same process applied to thinking and human relations, how I perceive and interact with people either in thinking or in conversations. I became aware through reflection, that people became what I was thinking or saying (distinguishing). I was living and interacting with them as if they were those perceptions. I knew in that moment I had hit on the core of objectivity-without-parenthesis, how the process of objectivity-without-parenthesis occurs, how distinctions in language changed in the emotioning of fear such that I was living them as being real and true in themselves without realizing how I was constructing them. This I later realised was the foundation of our social problems. Our problems in human relations are a result of our perceptions, what is going on in language and emotioning, how our perceptual behavioral patterns in thinking and relating co-ordinate around the emotioning.

We live an intrinsic perceptual behavioral separation from one another in objectivity-without-parenthesis that is based on the emotioning of fear. What will become apparent later on though is that we are not really separate at all. We are living an objective myth.

The Psychology of Objectivity-Without-Parenthesis: Patterns & Dynamics in Human Relations

A pattern is a repetition. Our emotioning reflects patterns of perception that appear in the dynamics of human relations or when we are thinking in solitude. For example, in the emotionings of fear, distinctions, perceptions of denial become apparent. These patterns of perception, distinctions, are associated with low self esteem and insecurity. Image and identity become important to the self. "I'm not beautiful enough." "I look terrible." "I'm too fat." We become very 'self-conscious' and appearance matters.

Other perceptual patterns of denial and low self-esteem are unworthiness and sometimes depression. For example, "I'm no good." "I'm a failure." "Nobody cares or loves me." "I don't feel like living anymore or, I feel like giving up." Feelings of being lost, alone and confused may also appear.

In the emotionings of anger and control the perceptual behavioral patterns of blame, guilt and submission show up. "It's my fault. I've done something wrong." What follows is generally guilt and appeasement, the desire to please or gain acceptance and approval. These perceptual behavioral patterns are also associated with insincerity. We will do or say things that we don't really mean to gain acceptance and approval or to appease someone in the dynamics of blame and guilt. As a result, doubt, suspicion and mistrust may appear for the person that we are interacting with in these kinds of dynamics.

These perceptual behavioral patterns reflect denial in the dynamics of human relations, usually experienced as a denial of the self. These perceptual behavioral patterns of denial only appear in the denial or suppression of emotionings as these patterns do not appear in the emotionings of love, trust and acceptance.

What is also interesting is that in the denial or suppression of emotionings, our experiences, realities etc., appear to be fixed, real and there in themselves. But how do the perceptions, experiences become fixed? When we suppress, we resist. What we resist persists, there is no change in the pattern. The pattern of our emotioning and distinguishing persists and experiences are lived as being fixed, there in themselves. For example, "I'm going to feel like this for the rest of my life." "My life is always going to be like this, it's never going to change." Or, "I will always be like this. I'm never going to change."

When we resist the experience, an attempt to change the pattern, the intensity of the experience along with the emotioning will increase and as the intensity does not feel comfortable, there is further resistance as we attempt to get out of the experience. The pattern however, will not change in resistance. The only way out is validation, validating the emotioning and what is showing up in terms of patterns of distinguishing (perceptions). Be the pattern.

Culturally, accepting and validating is difficult to do as resistance is familiar. However, in accepting or validating the tension (fear and resistance) in the body, the tension or intensity of emotioning eases and perceptual behavioral change occurs naturally.

The shift that takes place in the perceptual behavioral patterns of fear and control is not trivial in human relations nor in how we preceive ourselves and others as this is how we construct differences whether they be racial, gender, cultural and/or religious. Wars and conflict arise not to mention many of our other social problems.

Perceptual Differences

We are not the same in the world of differences. We are not human beings. We are objects separate from each other, individual selves living independently in our own little worlds, isolated and disconnected as we try to convince each other through rational arguments about how the 'world really is'. There is a separation between the self and the other, inherent differences that are based on our interpretations about what is there in reality. Thus, we are not living or sharing the same world. As a result, 'we' are not the same, we are not living in the same world together. It is not a world that is shared in mutual understanding. Instead, we are either denying or are afraid of the world of the 'other'.

Their world is not legitimate as we compare, assess and judge their world to ours.

In assessments and judgements we bring forth differences in distinctions of comparison. These comparisons are based on skin color, behavior, facial or bodily characteristics, dress or material objects. The 'other' appears to be different to 'me', the self. "He or she is better than me." "He or she is more beautiful." "What have they got that I don't have." "His or her skin color is different to mine." "They have more than I do."

When we see/distinguish that we look, think, behave or dress differently, other perceptual patterns can arise, specifically those associated with gender and race. For example, perceptions of superior/inferior often go hand in hand with intelligence, i.e., civilized and intelligent over uncivilized, barbaric, primitive, dumb, stupid and ignorant.

We only distinguish perceptual and behavioral differences in the emotioning of fear. Thus our perceptual behavioral patterns will reflect denial. Another human being is not legitimate or equal, because they are perceived and related to as being 'different'. Thus, they will be perceived to be a threat in the patterns of suspicion and mistrust. As a threat, the 'other' must be exterminated, gotten rid of. Violence and aggression appears, racial or religious wars may ensue.

Culturally, our worldview reflects what is going on in our emotioning and perceptual behavioral patterns both in thinking and in human relations. It is a worldview that is constructed or constituted in the explanatory path of objectivity-without-parenthesis that reinforces and supports separation as will be shown below.

The Worldview Of Objectivity-Without-Parenthesis

Separation: Perception

From the experiential reflections below we will see how separation is lived and experienced in an internal/external worldview from the explanatory path of objectivity-without-parenthesis.

Our worldview reflects that there is a pre-existing self or mind that is doing perception in the body. There is a subject perceiving what is going on in an external world. The external world determines what the subject sees through information. It is the mind that gathers and stores the information from the external world. The mind stores the information in its memory.

It is the self, the subject who is gathering and interpreting information from the mind about what is there in itself, in the 'real' or external world. Physical objects are assumed to determine what the mind sees through the sending and receiving of information – representations of the environment. These representations from the environment or external world that the mind receives, are also assumed to exist in the brain, in the head. We are walking around with objects, representations in our heads.

These representations of the external world, perceptions, are lived as an internal world. The internal world is that which the self interprets from the information that it receives from the mind. The self's experiences are based on these interpretations of information and are thus assumed to also exist in the head as representations. Furthermore, both the mind and the self control what the body does, behavior.

Behavior is also lived as being separate or independent from both the mind and the self and is assumed to exist within the body. To control behavior is to control the mind, reprogram it, give it new information so that the appropriate behavior happens or takes place.

As a result of this worldview, dynamics in human relations are perceived as instructive i.e., we assume that we are determining or controlling perception and behavior through the sending and receiving of information. And if there is no understanding taking place in human relations it is commonly believed to be a problem of interpretation, with how we are sending and receiving information and how we are interpreting what is received. Differences in the interpretations of experience arise and arguments may ensue about who is right or wrong about their interpretations of 'reality'.

Separation: Behavior

So how is behavior perceived culturally in an internal/external worldview from the explanatory path of objectivity-without-parenthesis? The self, the doer, needs a fixed reference point, a direction in order to know where it is going in life, thus a direction is set which is usually measured and assessed according to time. Having a direction and getting somewhere validates the existence of identity – the doer, the 'who'. The self or identity is in control of behavior and needs to know with certainty where it is going and what is going to happen, i.e., future behaviors. The self does this by establishing criteria, goals and outcomes, abstract concepts, based on perceived needs and wants. Notice how the perceptions of behavior become lived as being separate and external as if behavior existed independently from what we do.

Needs and wants are usually distinguished as a must or a should, what must or should happen. This is where demands and expectations appear in human relations. We force (control) our behaviors to fit with what we think should or must happen. We demand or expect our needs and wants to be fulfilled. Having our needs and wants met shows that we are in control. We can have and get what we want. We are in control of life.

Once the self has a direction, a plan, the goal or outcome is perceived as a fixed reference point for what the self wants or expects to happen, i.e., future behavior. The reference point becomes an 'it' out there somewhere (external reality) that the self is trying or determined to get to. The perception is that when the self reaches the goal or outcome, the self will be happy and fulfilled. However, happiness and fulfillment is short lived. So the self immediately begins to set criteria again in the endless search for happiness, love and fulfillment. The self is back in control.

These perceptions of the self and behavior are conserved in our daily living and are fraught with difficulties in human relations. For example, blame and denial appears when our perceived needs and wants (expectations) are not met, or when we perceive that someone is doing something that we think they shouldn't be doing. In other words, their behavior is not matching what we perceive or think should be happening.

Moreover, in these kinds of dynamics we feel pressured. Someone wants something from us. The pressure may be resisted, hence we will do forced co-operation, behavioral patterns of submission and insincerity, we are doing behaviors because we have to meet the criteria or expectations of others.

What happens when goals, plans, outcomes don't turn out the way that we want them to? Disappointment, pain, frustration shows up along with the perceptions (patterns of distinguishing) of "I'm a failure. I haven't achieved what I wanted to do." "There is something wrong with me." Or, "life isn't turning out the way that I want it to." We end up depressed and frustrated. Furthermore, there is a mismatch between what we think should be happening and what is actually happening. The moment does not fit with the criteria of what we want to happen, behavior.

These kinds of perceptual behavioral patterns both in thinking and in human relations throw us because we firmly believe that we are in control. We are convinced and certain that we know what is going to happen. In fact, knowing with certainty what is or isn't going to happen is very important for the validity of identity, because if the self is wrong, then the question of perception comes into play. As a result, the self may doubt and question its ability to perceive, to know or control what is going to happen, behavior. Confusion appears and the self doesn't know what is right or wrong anymore.

Aside from the psychological and emotional consequences in thinking and human relations, it is obvious that we live behavioral and perceptual uncertainty. Thus, the worldview of objectivity-without-parenthesis is inconsistent with our experiences.

Perceptual and Behavioral Uncertainty: Contradictions and Inconsistencies

There are many inconsistencies and contradictions between what we live as being real and true in objectivity-without-parenthesis and what we actually experience. This is due to the changes and shifts in our emotionings, hence the changes and shifts in our perceptual behavioral patterns. The fact that we live behavioral uncertainty, i.e., our criteria in terms of plans, goals and outcomes does not determine what is going to happen, behavior, shows we are not in control. Control is a myth in that sense. Moreover, how can an abstract concept, a word in language, do or control behavior? Perhaps it is not the self that is determining behavior after all, rather it is the biology.

We cannot change what is happening in the moment. What is happening is what is happening. In this way, we never do mistakes or errors in terms of inadequate behavior. Biologically, every behavior that we do is adequate.

A mistake or error occurs after the moment in which we reflect on what we did, usually as a judgement. The behavior does not fit with the criteria of what should be happening and recriminations usually appear. "You have made a mistake. What is wrong with you. Do it again and do it properly this time." Or, "There are mistakes here. This is not what I asked you to do. Didn't you listen the first time?"

We also live perceptual uncertainty i.e., we cannot distinguish in the moment between perception and illusion. For example, when we hear a noise we try to distinguish what the noise is and where it is coming from. We may distinguish that it is someone knocking on the door. We go and investigate and find that there is no one there. So we dismiss that perception as an illusion.

Illusions always occur a posteriori, after the moment in which we distinguish a perception. For example, we hear the noise again and go looking for the source, and find that it was a blind knocking against a window. So we dismiss the first perception as an illusion and validate the latter experience as a perception. From this example we can see that it is not the environment that is determining perception. Rather perception as a behavior is determined by what is happening in our structure, in our biology. That is, we hear what we hear according to what is happening in our biology. We hear a noise and then bring forth a distinction as a perception about what we heard.

The same situation applies to lying. It is very difficult to tell whether someone is being honest. We accept explanations and each other easily in the emotionings of love and trust until we find out that we have been lied to. As a result, all of the perceptions that we lived as being true or valid are now dismissed as illusions. Deception and mistrust appears. The dynamics of trust and acceptance are broken. Separation ensues.

These reflections however, are reflections about what occurs in our emotioning and how that influences our distinguishing in language both in thinking and in human relations. We may dismiss one perception in one moment as an illusion over another. The same applies to the validity of our perceptions in human relations i.e., we validate or invalidate perceptions, experiences etc., according to our explanations. As such, explanations and perceptions are not true in themselves. Rather, they are validated or invalidated through what we do in human relations according to our emotioning in any given moment i.e., we may or may not accept an explanation depending on our emotioning. For example, if we are in the emotioning of anger, then we are not likely to accept an explanation of why a person did this or that. However, if we are in the emotioning of love and acceptance, then we are more likely to accept an explanation.

Explanations are also accepted or not accepted according to our emotional preferences. We may like or choose one explanation over another because we like or dislike it.

In objectivity-without-parenthesis we live as if our interactions are instructive, as if we are determining or controlling perception and behavior through our talking. This is not the case however, as our daily experiences reflect. We ask, "What did you say? I didn't hear you, can you repeat that again?" Or, "I don't understand, what did you say again?" "You are not listening to me. That is not what I said. Please listen to what I'm saying." The same is true when we ask someone to do something and they do not follow through.

All of these examples clearly show that we are not sending or receiving information. We are not instructing or determining perception and behavior. We do not operate with inputs, information and outputs, behavior. There are no representations of the environment (information – objects, words, pictures, etc) floating around in the brain or the nervous system. Nor is there a mind in the brain, a self, a doer, or behavior for that matter residing in the body.

We are living an existential and objective myth in the explanatory path of objectivity-without-parenthesis because it is inconsistent with what is actually going on and what we experience. Moreover, as we will see, how we construct a world of separation, i.e., an internal/external world and what we live as a result, does not appear in the emotionings of love, trust and acceptance. Instead we construct a very different world, a world that is relational, congruent with the operation of our biology, how we are and what we do as living systems. This is the awareness of what shows up in the explanatory path of objectivity-in-parenthesis.

The Worldview of Objectivity-in-Parenthesis

In the emotionings of trust, love and acceptance there is no perceived separation because our distinctions in language do not become objectified as if they are entities in themselves, rather they are just part of the flow of distinguishing, either in thinking or in human relations. As such, we become aware that perception is taking place in our distinguishing in what we do, in interactions and thinking.

We do this irregardless of whether we are in the explanatory path of objectivity-without-parenthesis or objectivity-in-parenthesis, so this reflection is an awareness, what we become aware of when our emotioning shifts to love, trust and acceptance. We become aware of the relation, the doing, the interaction. The distinction relational languaging is another way of describing the languaging process, what shows up in the emotionings of love, trust and acceptance.

When we co-construct our experiences, worlds, realities etc., in patterns of distinguishing, we are bringing forth together or co-constructing a shared or relational world – there is no separation. Experiences are constructed in the happening of distinguishing in language without a doer or perceiver. This does not mean that the self as a distinction in language does not appear. It means that the self just remains a distinction in language and does not become objectified as an entity in itself that is doing perception. The self in this manner of distinguishing is lived and experienced in the relation in language as a construct not as an entity in itself.

How do we know though, that we are doing behaviors of distinguishing in emotionings of trust, love and acceptance? We can tell by how our bodies feel. There is no tension as there is in resistance and fear. Our bodies feels calm, relaxed. We feel good. Alternatively, we can also feel passionate, excited, inspired etc., we feel great.

In these emotionings, what we commonly distinguish as trust or acceptance, the plasticity of the nervous system is expanded. Thoughts, behaviors of distinguishing in language, appear as if out of nowhere. These distinctions can also trigger inspiration, passion and excitement. There is a knowing of what to do and the doing flows naturally. We are caught up in the doing. The doer is not there, nor is the self, nor is time. There is just the doing.

Sometimes there is a feeling of urgency in the doing. There is a knowing that what is showing up in the doing is important. This knowing appears as intuition or instinct, biological dispositions, what we distinguish as hunches or gut feelings.

More often than not it is difficult to explain or know how we know. We just do. There is no reference point in terms of the self or mind anymore. The reference point is the emotioning in which knowing appears as a biological disposition – the biology of knowing. Knowing appears in two domains – in the domain of doing, distinguishing and also in the domain of the biology – biological dispositions or what we distinguish as feelings.

The biology of knowing is also triggered in interactions. For example, when we relate in the dispositions of trust, love and acceptance, the person that we are interacting with says something that triggers particular kinds of biological dispositions associated with knowing. Examples from my experiences are goosebumps that appear on the skin, a tingling that goes up and down my spine or crying that happens spontaneously.

These kinds of knowing appear as an understanding that can sometimes go beyond words or even explanations. In other words we know the understanding is true or right without even knowing why or how. The knowing appears as an expanded state of awareness, a deep, shared and mutual understanding.

Relational Languaging: The Constitution of the Relational World

When we construct or bring forth our experiences, explanations and realities from the emotionings of love, trust and acceptance, we begin to live or experience a different world, a world of expansion, a constructed world that reflects the expansion of our perceptual behavioral potential. This is where intelligence, wisdom and humanness appears; the relations of honesty, respect, empathy, compassion and mutual understanding.

In the human relations of openness, honesty, trust and acceptance we can experience a mutual understanding that is beyond words, i.e., the understanding occurs in how we are co-ordinating together in the emotioning, patterns of co-ordinations of co-ordinations of behavior. We can experience or distinguish a human being as knowing them. There is a knowing or sense of familiarity about the person, like you've known them before. It's like you KNOW each other without getting to KNOW each other. There is a depth of understanding without words.

There is also a sense of just being in which time does not appear. The self is not there, nor is needs, wants, expectations, judgements, goals, outcomes etc. There is just a mutual desire to BE together. Whatever happens as a result, does not matter. There are no demands or expectations about what will or won't occur. These are what I distinguish as the relations of love, acceptance and caring without a self or identity as neither appear here.

In the relations of love, acceptance and caring first thoughts or intuitive distinctions show up. We are living and co-constructing a shared relational world in mutual understanding, in the relations of openness, honesty and sincerity. We are relating in relational knowing where passion, excitement, and other intuitive dispositions such as goosebumps or crying appear in an expansion of shared awareness.

We are living and experiencing together another world where we are not separate individuals caught up in the identity of self, race or gender. We are together as equals, as human beings. Gender, race, or cultural differences do not appear in relational languaging, the co-construction of the relational world.

A shared world can also happen in different ways but in the same flow of emotioning. For example, if we reflect on the dynamics between musicians and an audience we can see that understanding does not just appear in words, but in the rhythms or sounds appearing as co-ordinations of co-ordinations of patterns of behavior, the bodies of the audience move automatically with the rhythm of the music. There is a shared and mutual understanding in the consensual flow of co-ordinations of co-ordinations of actions, of shared languaging rhythmical patterns of behavior.

Other kinds of relational experiences are relational visions, premonitions, knowing something is going to happen before it does, dreams that are remembered upon waking up, and synchronistic happenings. Relational visions appear as if from nowhere in the relational space. Premonitions are usually experienced as instinctive feelings along with intuitive distinctions. Premonitions are also associated with relational visions. Experiences of deja vue happen as if from nowhere too, as if the experience has been lived before, perhaps in a dream.

The dreams are also quite different to that of normal dreams. There is a knowing, usually experienced as a feeling of importance. And quite often what I experienced in the dream will appear in daily life as a synchronicity of events i.e., I will meet the person that I saw in the dream, or a particular situation will happen that triggers the remembering of the dream. Other kinds of synchronicities are: I will be thinking about someone and then the person calls me, I will be thinking of a particular situation and it will just happen, again as if from nowhere.

Time is not relevant as events or situations just happen. Whatever happens is whatever happens. It just is. There is trust and acceptance in the unfolding of living. There is no effort, trying or forcing to make behaviors happen as I 'want' them to. Wants, needs, expectations, goals, outcomes, criteria etc., are no longer important or relevant. Demands in terms of forcing or controlling behaviors associated with the self or others, are also not present in the relational world of living.

Instead of using time to co-ordinate my behavior, i.e., wants associated with fulfilling goals and outcomes, I pay attention to what is showing up biologically i.e., intuitive or instinctive feelings or intuitive distinctions – the biology of knowing or cognition (distinguishing). These intuitive feelings can be what does or doesn't feel comfortable. If my body does not feel comfortable I will not explore a particular line of thinking or be in particular situations or interactions. If my body feels comfortable and a knowing, passion or excitement appears, or something feels right, then I follow that. There is a flow to the happening. I have no idea of what is or isn't going to happen in this flow or where it is going to lead me so to speak. Having a reference point, an end point in time or the ending of something, in terms of behavioral certainty is not relevant. It is not even an issue. It does not matter. What matters is the doing and what shows up in the doing, the sharing, learning and understanding.

Identity associated with the self is not there in relational languaging as mentioned previously. Thus, issues, or patterns of perception associated with an objectified self, i.e., confidence, beauty, acceptability, whether we are loved or accepted or not, no longer is an issue. When these issues or patterns of perception do show up, either alone in terms of thinking or in an interaction with other human beings, then the core of these perceptions/behaviors are arising from fear. As it is only in the emotioning of fear, mistrust and control that these patterns arise.

Moreover, many of the problems or issues associated with fear, mistrust and control in the normal dynamics of human relations do not appear either. There is no perceived 'other' who we feel afraid of or threatened by that appears in fear and mistrust. Nor is there the perceived endemic differences associated with racial and gender identity, rather there is a sense or understanding of sameness. We are not the same, and yet we are the same. We are relating as if we are the same. There are no differences. We are human beings. We are relating and interacting as human beings in a world that is shared and understood, where we are learning, sharing, experiencing and helping each other in natural co-operation. As human beings we are distinguishing each other as being the same in the validation and legitimacy of both our biology and behavior – the biological and behavioral relations of love, trust and acceptance.

By paying attention to our emotioning, we can now see what does and doesn't show up in the perceptual behavioral patterns of human relations. We can know and understand the worlds we are co-constructing and maintaining in our patterns of distinguishing. That is, the emotionings of fear, mistrust, resistance and control, the suppression of emotionings or love, trust and acceptance.

This shift in the emotionings constitutes a paradigm shift. A shift in our perceptual behavioral patterns in which we bring forth, constitute, two worlds as a manner of living and relating. The question is however, how do we live and maintain the shift from objectivity-without-parenthesis to objectivity-in-parenthesis?

Moving Between the Two Worlds with Ease: Emotions

To move from fear into trust, we can share what is showing up. Sharing what is showing up in the moment validates our emotioning (biological dispositions) and the perceptual behavioral patterns that we do in thinking and human relations (interactions).

For example, if we are suppressing our emotioning and fear shows up as a result, all we need to do is to validate the fear. In validating, the pattern of emotioning changes automatically along with the associated perceptual behavioral patterns.

To not validate or accept what is showing up, the perceptual behavioral patterns of resistance will persist. The world, as we live, experience and distinguish it, will appear to be fixed and in itself. The world of objectivity-without-parenthesis will be maintained and conserved through resistance. Moreover, forcing a change in the emotioning won't work either. Instead, resistance will be reinforced and we will remain in the world of objectivity-without-parenthesis.

Reflecting on what is showing up is another way of validating. Reflecting frees the desire for certainty or control. We can reflect on what is happening in the dynamic, interactions, or emotioning and learn and understand why or how this particular emotioning or dynamic is showing up.

Reflecting reminds us that we are only triggering changes in the biology and interactions. We are not controlling or determining perception and behavior in terms of instructive interactions, i.e., sending or receiving information. We are not making other people feel a certain way, hence we need not feel guilty or responsible for their experience. Their experience, what they distinguish about what is happening, depends on their emotioning. From this understanding, we can explore what is happening in the dynamics of biology and behavior, the two domains of existence. So instead of blame, "you made me feel this way, it's all your fault," we can ask – what is being triggered? Or, what is happening for you? Please share what is showing up.

These kinds of questions automatically invite a reflection on what is happening in the biology (I felt this and that) and what is happening in the domain of interactions (I did or said this and that as a result). Or, when you did this and that, that triggered these feelings which triggered those perceptual behavioral patterns.

In this way, the dynamics of our perceptual behavioral patterns are being explored without blame, assessment or judgement. Rather we explore together in curiosity, learning and understanding of what happened, in the relations of love, trust, openness and honesty and as we do so, we move into the awareness of a relational world, the experience and behaviors of living and interacting in equality.

We can finally live in freedom and equality as one race, the human race where our perceptual behavioral potential is realised in the biology of cognition. We can live in unity in helping, caring, respecting and supporting each other in realising our dreams, passions, intuitive or instinctive desires. The possibilities for what we can accomplish together as a unified race are endless.

CHAPTER 5

THE CONSTITUTION OF CULTURES

Overview: The Patriarchal and the Matristic

What is a culture? How is a culture constituted or constructed? A culture is constituted or constructed in human relations as a network of conversations – a braiding of languaging and emotioning (Maturana, 1988). Our languaging, both verbal and non-verbal, follows the flow of our emotioning, biological dispositions.

In languaging, the doing of language, we distinguish each other, physical objects and conceptual objects in co-ordinations of co-ordinations of distinguishing, in other words, patterns or repetitions of behavior. In the process of distinguishing we create perceptions, experiences, explanations, descriptions, realities, worlds and, depending on the flow of our emotioning, the distinctions in which we bring forth, create, perceptions, experiences, realities, worlds, etc., become entities in themselves either as a fixed reality, or just part of the flow of distinguishing, a co-constructed reality. See Chapter 4, The Construction of Worlds.

Understanding this process of distinguishing in language, languaging, as perceptual behavioral patterns is not trivial. It is an awareness of:

- How we are bringing forth and living our perceptions as experiences, worlds, realities etc.;
- How our perceptions, experiences, realities, worlds etc., and behaviors emulate our emotioning, i.e., the dynamics, perceptual behavioral patterns of human relations and thinking.

The world as we live and know it according to our conversations not only constitutes our manner of living and relating, but reflects it also. The same applies to culture. That is, culture as a network of conversations shows how the world is conserved as a particular manner of living and relating. It is a circular process.

Understanding culture as a process is important. It enables us to see how our worldview, manner of living and relating is created and maintained in the many perceptual behavioral patterns that we do in conversations, human relations and, more importantly, we can see how these perceptual behavioral patterns appear or change according to the flow of our emotioning – how we constitute and live the worlds of objectivity-without-parenthesis and objectivity-in-parenthesis.

In the explanatory path of objectivity-without-parenthesis our manner of living and relating reflects a patriarchal culture that is defined by the emotionings of fear, mistrust and control. In the explanatory path of objectivity-in-parenthesis, our manner of living and relating reflects a matristic culture that is defined by the emotionings of trust, love and acceptance. The term 'matristic' refers to relations of equality, a manner of living reflecting a socially constructed relational, circular/systemic worldview.

The Patriarchal Culture as a Manner of Living and Relating

The emotions that define our relations in the explanatory path of objectivity-without-parenthesis are fear, control, mistrust, domination and appropriation (ownership). In these perceptual behavioral patterns we are objects, because what is important is getting what we want. Getting what we want assumes control, the certainty of knowing that we are in charge of our survival and existence.

This process of objectification can be seen in several domains of conversations, perceptual behavioral patterns in daily life.

Business: Organizations and Customers

In business the general perceptual behavioral patterns that are conserved in conversations in relation to customers are:- customers are there to satisfy the needs or goals of the organization as economic objects. Management/staff talk about customers as sales quotas to be filled for the month that are translated into sales figures and profit margins, economic statistics. As such organizations do not serve their customers. Customers serve the needs and goals of organizations. Needs and goals obscure service.

Staff/management lie or deceive customers about their products. That is, generally the product is not made by the organization, but is made by someone else even though the organization's label or brand is put on the product. In buying the product, customers assume that they are buying a reputable and quality brand based on the organization's reputation. They are however unaware of who has actually made the product thus calling into question product quality.

As many organizations are cost cutting these days in order to make more profits, many of their services are out-sourced to other organizations or companies in developing or third world countries. Customers are deceived into thinking that they are dealing with the company in relation to a particular product or service, when they are in fact dealing with another company.

Out-sourcing and cost cutting is also attributable to the exploitation of people in third world countries as low wages and poor working conditions are maintained for higher profit margins. Concern for the well being and living conditions of these people is not taken into consideration.

Business, Politics & Globalization: The Core Perceptions & Behavioral Ramifications

The conversations that are maintained as perceptual behavioral patterns in politics are generally about economic rationalism and globalization. That is, the pursuit of material objects resulting in economic wealth and power – the power to dominate and control human behavior. Citizens are perceived as objects to fulfill the personal or hidden agendas of politicians.

Political leaders who subscribe to economic rationalism and globalization rarely care about the people they supposedly represent. They are interested in power; the acquiring of material wealth and the associated status and prestige that goes with it. As a result, leaders become blind. The needs or concerns of the people are not seen as the acquiring or getting of material objects obscures the perceptual behavioral relations of acceptance and caring. These perceptual behavioral patterns are also relevant to business and many other areas of daily life.

The perceptions of materialism, the acquiring and owning of objects are becoming more wide spread, particularly in what we call developing or third world countries where material objects are marketed as status symbols. That is, if a person owns this particular object or status symbol, he or she will be perceived in high esteem among their community or society at large. They are perceived as having social importance based on the acquiring of a status symbol.

Little thought however, is given to the consequences. Jealousy may be triggered resulting in crime and theft.

The association between identity, and the ability to acquire and own material objects as status symbols is not a trivial one as it generates a division between the haves and the have nots – the rich and the poor. This division among people only exists in the acquiring and ownership of material objects.

To emphasize this point, I was conversing with a stone artist from the Shona Tribe in Harare, Zimbabwe, Africa. He was sharing with me how the perceptions are changing from the traditional ways to the western ways. He was saying that someone who has no money or few material objects is perceived as being nothing, a worthless human being with no identity, ostracized by the tribe or community based on what they do or don't have.

This perception of identity based on materialism is very different to the traditional or tribal way of life. That is, identity is not defined by objects, materialism, but by the social and family relations of the tribe. He or she belongs to something much larger than his or herself, as an extension of the family and their ancestors.

Also in the traditional way of living, there is no importance given to objects other than to help them and assist them with the preparation and storage of food for example. Objects then, are not owned. Thus, there is no rich and poor. Objects are shared among the family or tribe. As a tribe they help and share among each other, a mutual co-existence. As such, their identity and survival depends on each other and the environment in which they live.

In Harare, the ramifications of westernization are taking their toll on the traditional way of life. Social and family structures are breaking down. There is a higher consumption of alcohol among men resulting in violent and aggressive behavior. There is more abuse of women and children than there was before and, there is a high crime rate in relation to theft as people struggle to exist and survive in the world of materialism.

Material wealth can not only defines our identity, but also our happiness, existence and survival. We never have enough. We need more. The more we have the more secure we will feel in an uncertain world where anything can happen.

Getting and having what we want, material objects, are the keys to success, happiness and fulfillment in objectivity-without-parenthesis. Happiness and fulfillment depends on acquiring and owning material objects, i.e., the more we have the happier we are supposed to be. We are never truly happy or fulfilled though. Happiness and fulfillment is short lived. So we keep on looking, repeating the pattern of going after and getting what we want, trying to reach the ultimate states of happiness and fulfillment and when those states are reached, we start looking again.

Conclusions: Why a Paradigm Shift is Needed

The consequences of objectivity-without-parenthesis are obvious in human relations. We remain blind and indifferent to each other in the pursuit of materialism – the never ending quest for economic growth, the ownership and appropriation of objects. So endemic is this quest, we lie, deceive, manipulate, dominate and control each other to get and have what we want.

Moreover, objectivity-without-parenthesis is a paradigm that supports separation and inequality, not unity. We are separated by the pursuit of materialism and economic rationalism in which the gap between rich and poor arises and, we are separated by perceived inherent differences that are associated with identity. In this way our identity is not human. Our identity is categorized hierarchically according to the relations of superior/inferior.

In gender relations men are superior to women. In race relations the white race is more superior to the black race, or the Western culture is more superior to that of indigenous or tribal cultures. In religious relations Jews, Christians, Muslims, Protestants etc., try to prove who is more superior, usually through conflict. As a social construction, we believe that we are those identities.

It is not our beliefs that separate us. It is our perceptual behavioral patterns in human relations that is the problem, because in these perceptions we objectify each other. We are seeing, perceiving each other as objectified identities that are based on fixed and inherent differences. The differences however, are only what we distinguish them to be. Differences are a social construction.

These perceptual behavioral patterns also influence the perceptions of our behavioral potential – intelligence. For example, gender and race. Women were or are considered to be less intelligent than men, perceived as being incapable of making important decisions or doing the same roles or tasks as men. The white race is superior and more intelligent than the black race. Black people, African or African/American, are perceived to be uneducated, dumb, stupid and lazy, incapable of being intelligent, sub-human. They were relegated to serving their white masters in subservient roles, accepting that they were not intelligent enough or capable of doing more. Thus, there was a perceived limit of their behavioral potential. These perceptions are also relevant to indigenous and poor people often resulting in exploitation.

Of course many of these perceptions have proven to be a myth. But in the example of gender and race, many women, Africans and African/Americans still struggle with the consequences of superior/inferior relations trying to overcome their insecurities about their intelligence and behavioral potential.

Our Crisis & Our Possibility

Our crisis is a human one. The paradigm of objectivity-without-parenthesis is not working. Our social and environmental problems exist as a result of our perceptual behavioral patterns in thinking and human relations. We, as a species, have constructed and maintained these patterns socially and culturally as a network of conversations and, if we do not question and reflect on these patterns and the emotionings in which they arise, the patriarchal culture will continue. The patterns that constitute the social construction of the patriarchal culture in human relations will continue.

It is obvious that a shift in paradigms is needed if we are truly committed to resolving the many problems that we have created for ourselves. This shift I feel is a natural solution, because the problems of objectivity-without-parenthesis do not exist in a relational worldview as I have already mentioned. Moreover, this worldview is reflected in a matristic manner of living and relating – a matristic culture.

The Matristic Culture as a Manner of Living and Relating

A matristic culture reflects another kind of manner of living, relating and worldview that appears in the emotionings of love, trust and acceptance. The experiential awareness of objectivity-in-parenthesis, the relational world that reflects how we are as biological living systems. Our worldview, manner of relating and living is consistent or congruent with how we are.

We live expansions of awareness. Our systemic interconnectedness, both socially and environmentally becomes more apparent. Several other phenomena show up in the emotionings of trust, love and acceptance. Our intuition and instinct become stronger. We know what to do. We have strength, clarity, vision and purpose. We know what we want to do in life. The doing is natural not forced. We experience ease and flow. There is no effort in doing what we do, it occurs naturally. We are immersed and consumed in doing what we are doing. We experience effortlessness.

More often than not, our behavioral potential is expanded when we are immersed in something that we love to do or feel passionate about. We go beyond what is considered to be our normal perceptual behavioral patterns. We make great leaps in our understanding or accomplish behavioral feats that we normally consider as being impossible, wondering how we did it.

Working takes on a different perspective. Tasks are accomplished in a much shorter time. Productivity and efficiency is increased, as there is a genuine care and concern for how well the task is being done.

There are also experiences of unity and belongingness. For example, I went to the Olympic Games held here in Sydney in 2000. The city, which is usually very hectic, cold and indifferent, suddenly became alive with people that were very friendly, happy, relaxed and helpful.

The service in shops, restaurants, public transport etc., was remarkable as people were offering to assist and help in any way that they could and there was natural cooperation between staff and customers. On public transport people were talking to each other. This does not normally happen. The crime rate dropped substantially. The police, not having much to do as the crowds were very well behaved, stood around and talked to people.

I also spoke with a couple of volunteers who were driving officials and athletes around and they shared with me their experiences of not wanting to go home after they had completed their shift. They loved what they were doing and didn't want it to end. They didn't care about the long working hours or the fact that they were not getting paid. They were part of something unique, an experience of a lifetime, helping to make these games the best and friendliest ever.

The support of the crowd who cheered and encouraged the athletes to do their best, the volunteers and staff, all participated together in a feeling of unity. There was a common bond that united us as human beings. It was I feel, the desire to help, assist and support each other in realizing a dream. A dream that was realized in natural and spontaneous cooperation that made these games the best ever.

Many people felt a sense of loss and depression when the games were over. Many remarked why it could not be like this all of the time. Something had occurred that was special and now it was missing. The transformation that took place in many people's daily lives was only temporary but maybe this transformation was enough to show us what is possible when we come together in unity to achieve and realize a shared and common dream.

Of course there are many other examples of this kind. Usually it takes some kind of crisis like a famine, flood or other kind of disaster to unite us as human beings; we all help and pull together to accomplish something that is much larger than our individual needs, wants, etc. Racial, gender, cultural and religious differences do not seem to matter anymore. We are together as one, the realization of our systemic interconnectedness and humanity.

These examples and experiential understandings show that a matristic manner of living and relating is possible. The matristic worldview defined by the relations (perceptual behavioral patterns) of loving, trusting, caring and acceptance, is not about acquiring and owning material objects. This is not to say that we don't need objects because we do. Objects just support or sustain our way of life.

Identity, status and prestige, all of the perceptions of objectivity-without-parenthesis laid out in Chapter 4, The Construction of Worlds and partially in this chapter, do not appear here, rather, there is an overwhelming sense and desire to help and support others in the passion and knowing of doing what we love, something that we do naturally, without effort or force. Goals, outcomes, plans etc., all the typical reference points of objectivity-without-parenthesis in relation to behavior, do not appear here. There is no sense of direction, i.e., where, when and how.

No beginning or end point, just a natural unfolding of a sequence of events.

Many indigenous cultures were matristic. Their worldview and way of life reflected a systemic understanding of themselves and nature, of how the biosphere worked. This was and has been part of our human history.

Perhaps through the understandings laid out in this chapter and the rest of the book, we can reclaim this essential knowing of ourselves, a knowing that helps us realize that we are not alone. We are not isolated, separate individuals. We are and always have been part of something that is much larger than ourselves. We are very much interconnected in the spiral pattern of life, the heritage of our mutual co-existence, the essence of what it is to be human.

REFERENCES

1. Cull, J. (2000). *The Circularity of Living Systems – The Movement & Direction of Behavior.* The Journal of Applied Systems Studies, Cambridge International Science Publishing, Cambridge, England.
2. Maturana, Humberto R. (1998). *Reality: The Search for Objectivity or the Question for a Compelling Argument.* The Irish Journal of Psychology, 1988, 9, 1, 25–82. University of Chile, Santiago, Chile.
3. Cull, J. Gonzalez F. (1998). *Maturana Seminar*, Boston, Massachusetts. Partial transcriptions from audio tapes.
4. Maturana, Humberto R. Cull, J. (1997) *Conversation with Humberto Maturana, on objects, perception and internal/external world.* Transcript of taped conversation. Santiago, Chile.
5. Maturana, Humberto R. Mpodozis, J. (1997) *Perception: Behavioral Configuration of the Object.* Edited Version in English, 1997, University of Santiago, Chile.

6. Maturana, Humberto R. *Everything Is Said by an Observer.* In W. I. Thompson (ed.), Gaia: A Way of Knowing, Hudson NY: Lindisfarne Press, 1987, pp. 65–82.

GLOSSARY OF TERMS

Autopoiesis – Auto: Self or automatic. Poiesis: Production. Autopoiesis refers to cellular and molecular self or automatic production.

Bring Forth – Constitute, create, construct experiences in language.

Distinguishing – Distinguishing takes place in behavior. We do behaviors of distinguishing in language. We constitute words, thoughts etc., in behaviors of distinguishing in language.

Emotioning – Emotioning is the doing of emotions. The understanding of how our behavior changes or flows according to the varying changes taking place in our biology, i.e., the structure. These changes are distinguished as biological dispositions.

Experience – What we distinguish or bring forth in language as happening to us. That is, we constitute our experiences in distinctions in language. Words, thoughts, are brought forth in this way.

Generative Mechanism – A process that explains the experience or phenomena to be explained.

Manner of Living – A history of interactions that are conserved in the doing of living in recurrent interactions – patterns of perception and behavior.

Objectivity – Existence of objects, conceptual and physical, that are lived or experienced as being independent of what the observer does.

Objectivity-in-Parenthesis – Existence of objects depends on what the observer does. Objects, the naming of, both conceptual and physical, appear in distinctions in language.

Observing – Observing is distinguishing objects, conceptual and physical, in language. As such the process of observing takes place in behaviors of distinguishing.

Ontogeny – Ontogeny refers to a history of interactions.

Ontology – The study of existence.

Organization – Organization refers to the relations among the components of a system.

Praxis – The doing of living.

Relational – Relational refers to the domain of interactions, the doing of distinguishing and behaviors.

Structure – Structure refers to the components plus the relations of a system.

Structural Determinism – Structural determinism refers to the operation of a living system and how it is able to do behavior as a result of that operation.

Structural Coupling – Mutual change and adaptation between a living system and the environment in recurrent interactions.

Triggering – Our interactions with the environment or each other are not instructive. Hence we do not cause or determine what happens in the structure of our bodies, we only trigger or modulate changes.

INDEX

A

Acceptance 14, 15, 16, 37, 41, 42, 46, 50, 51, 60, 61, 62, 63, 64, 65, 66, 67, 69, 75, 77, 82, 83, 85
Autopoiesis 20–22, 89
Awareness 2, 4, 8, 14, 16, 17, 25, 26, 37, 41, 42, 45, 46, 47, 62, 64, 66, 71, 74, 82

B

Behavior 2, 7, 8, 9, 12, 14, 15, 16, 21, 23, 26, 27, 29, 30, 31, 32–35, 36, 37, 38, 39, 40, 43, 44, 45, 46, 48, 49, 50, 51, 52, 53, 55–58, 58–62, 63, 65, 66, 67, 68, 69, 70, 71, 72, 73, 74, 75, 77–79, 80, 81, 82, 83
Behavioral change 15, 31, 37–38, 52
Boundary xi, 21, 24, 30, 31
Brain 39, 54, 61

C

Certainty 12, 13, 14, 37, 56, 58, 68, 70, 75

Change & adaptation, *see also* structural coupling xi, 30, 31, 33, 36, 37, 45, 91
Co-existence 8, 16, 17, 29, 32, 41, 42, 44, 79, 86
Control 12, 13, 14, 15, 16, 23, 26, 37, 38, 48, 50, 52, 55, 56, 57, 58, 59, 61, 67, 68, 69, 70, 71, 75, 77, 80
Construction of worlds, realities, Experiences xi, 3, 4, 8, 9, 12, 16, 26, 43–72, 73, 74, 75, 80, 82, 85, 89
Cull, Jane 87, 95

D

Denial, *see also* suppression 14, 16, 41, 42, 50, 51, 53, 57

E

Emotioning, emotions xi, 2, 4, 9, 11–17, 23, 25, 37, 38, 43, 44, 45, 46, 47, 48, 49, 50, 51, 52, 53, 58, 60, 61, 62, 63, 64, 65, 66, 68, 69, 70, 71, 73, 74, 75, 82, 83, 89

Explanations 2, 3, 4, 6, 7, 8, 9, 10, 12, 13, 15, 16, 17, 48, 60, 61, 64, 65, 73
Explanatory paths 2, 11–17

F

Fear 23, 37, 38, 45, 46, 47, 48, 49, 50, 52, 53, 63, 68, 69, 70, 75
Function & purpose, *see also* Systems 28

G

Generative mechanism 5, 6, 7, 90
Gonzalez, F xiii, 87

H

Humanness 41, 65
Human relations xi, xii, 3, 8, 9, 12, 15, 23, 26, 27, 33, 36, 39, 43, 44, 45, 47, 49, 50, 51, 52, 53, 55, 56, 57, 58, 60, 62, 65, 69, 70, 73, 74, 80, 82

I

Identity 13, 22, 23, 50, 56, 58, 66, 68, 69, 78, 79, 80, 85
Information 32, 33, 39, 40, 54, 55, 61, 71
Interactions 4, 20, 23, 29, 30, 31, 32–35, 36, 37, 38, 39, 40, 41, 42, 44, 45, 61, 62, 64, 68, 70, 71, 90, 91
Interconnectedness 32, 82, 85
Interdependency 32
Internal/external world 46, 47–50, 54, 55, 62, 87

L

Language/languaging 2, 3, 4, 7, 8, 9, 11, 12, 14, 15, 22, 23, 33, 34, 36, 37, 39–40, 41, 43, 44, 45, 46, 47, 48, 49, 59, 60, 62, 63, 65, 66, 68, 73, 74, 89, 90
Learning 35–37, 68, 69, 71
Living Systems xi, 2, 17, 19–28, 29–42, 45, 47, 62, 82, 87, 95
Love v, xiii, 14, 15, 16, 41, 42, 46, 50, 51, 57, 60, 61, 62, 63, 64, 65, 66, 68, 69, 71, 75, 82, 83, 84, 85

M

Matristic culture 75, 82–86
Maturana, Humberto v, xi, xiv, 6, 12, 16, 17, 43, 73, 87, 88, 95
Mind 2, 5, 48, 54, 55, 61, 64
Mpodozis, J 87
Multiverse 3, 4, 8, 9, 11

N

Nervous System 15, 27–28, 37, 38, 61, 63

O

Objectivity-without-parenthesis 1, 2, 3, 5, 7, 13, 14, 15, 17, 41, 43, 44, 45, 46, 47, 49, 50–54, 54–58, 61, 62, 70, 75, 80, 85
Objectivity-in-parenthesis 1, 2, 3, 4, 5, 7, 14, 15, 17, 43, 44, 46, 47, 62–64, 70, 75, 82, 90
Objects 1, 2, 8, 9, 10, 15, 39, 44, 45, 46, 47, 48, 52, 54, 61, 73, 75, 87, 90
Observer 1–17, 20, 22, 28, 34, 39, 45, 88
Ontogeny 90
Ontological diagram of the Observer 1

Ontology 1, 3, 4, 90
Organization & structure, systems 22–24

P
Paradigm shift 44, 69, 80–81
Patriarchal culture 75–79, 82
Perception xi, 2, 14, 25, 26, 43, 44, 46, 48, 49, 50, 51, 52, 53, 54–55, 56, 57, 58, 59, 60, 61, 62, 63, 68, 71, 73, 74, 77, 78, 80, 81, 85, 87, 90, 95
Perception & illusion 26, 59
Perceptual differences 52–54

R
Recurrent interactions 31, 35–37, 38, 39, 45, 90, 91
Relational languaging 44, 62, 65–70
Relational world, constitution of 65–70
Reality 2, 3, 4, 8, 9, 11, 12, 13, 14, 39, 46, 48, 52, 55, 56, 74, 87
Resistance 38, 51, 52, 63, 69, 70

S
Scientific explanations 10–11
Separation 35, 49, 52, 54–58, 60, 62, 63, 80

Social systems 23
Structural coupling, *see also* change & adaptation xi, 30, 31, 36, 45, 91
Structural determinism 24–26, 91
Suppression, *see also* denial 51, 69
Systems
    characterization of, autopoeisis 20–22
    simple and composite unities 19–20
    function & purpose, *see also* function and purpose 28
    organization & structure 22–24

T
Thinking 3, 9, 12, 15, 37, 39, 43, 44, 45, 46, 47, 48, 49, 50, 53, 58, 60, 62, 67, 68, 70, 74, 76, 82
Triggering 21, 25, 32–35, 71, 91
Trust 14, 15, 37, 42, 46, 51, 60, 62, 63, 64, 65, 67, 69, 70, 71, 75, 82, 83, 85

U
Uncertainty 26, 58–62

V
Varela, Francisco xi, 17, 95

ABOUT THE AUTHOR

Jane Cull is the Founder and Lead Consultant of Life's Natural Solutions. Since 1994, she has developed a range and depth of expertise based on extensive research into the theories of living systems, the work of world-renowned Chilean Systems Biologists, Dr Humberto Maturana and the late Dr Francisco Varela. She has embodied and expanded on their work to provide an experiential and conceptual understanding of biology, human perception, the circularity of life and how we make the shift to construct a more sustainable world. Jane was also the Associate Editor of the Journal of Applied Systems Studies, a bi-annual academic journal dedicated to applied systems theory and is part of the Great Transition Initiative (GTI), an international network of scholars and activists that analyzes alternative scenarios and charts a path to a hopeful future. She is also on the Advisory Board of the Barcelona Consensus, a project dedicated to transitioning to a sustainable society.

Made in the USA
Lexington, KY
14 June 2015